The Case of the Missing Class Pet

by Thomas K. and Heather Adamson • Illustrated by Charlie Alder

The kids of Lincoln Elementary are on the case! When things go missing, they use math to solve the mysteries.

You, the reader, can help. As you read this story, solve the math problems. Choose your answer from the bottom of the page, and skip to the page for that answer. If you're right, you're one clue closer to solving the mystery! If you're wrong, you can try again.

What do YOU think happened to the class pet?

Start on page 3. ▸

Tuesday started out as a regular day in math class. Mr. Jeb was teaching a lesson about measurement.

Suddenly, Ella waved her hand. “Mr. Jeb, Lucky is gone!”

"His door must not have been completely latched!" says Juan.

Mark says, "Let's look for clues. Rats can hide in lots of places."

"Let's split up," Mr. Jeb says. "The room is 20 feet wide. Two groups can each cover half to search faster." How much of the room will each group search?

If each group will cover 10 feet, turn to page 10. ▸

If each group will cover 12 feet, turn to page 14. ▸

Think again.

A yardstick equals 3 feet. If 3 shelves equal 3 feet, there are 3 shelves left to measure. Those 3 shelves also equal 3 feet.

$$3 + 3 = 6$$

Turn to page 20.

Ella says, "We know that 12 inches equals one foot. So if it's a little more than that, we should measure in inches."

"Yes," Mr. Jeb says. "And what's the best tool to use to measure something that's probably more than one foot?"

If they use a ruler, turn to page 9. ▶

If they use a yardstick, turn to page 15. ▶

Think again.

It's 43 centimeters from the table to the shelf. But the chart showed Lucky's growth in inches. Use the same units to compare measurements. Measure with the inches side of the yardstick.

Turn to page 12.

Think again.

Twelve inches equals one foot. Because the shelf is a little higher than that, inches will be the most accurate. Inches and centimeters work better for short lengths. Feet and meters are more useful for medium distances, like the length of a room.

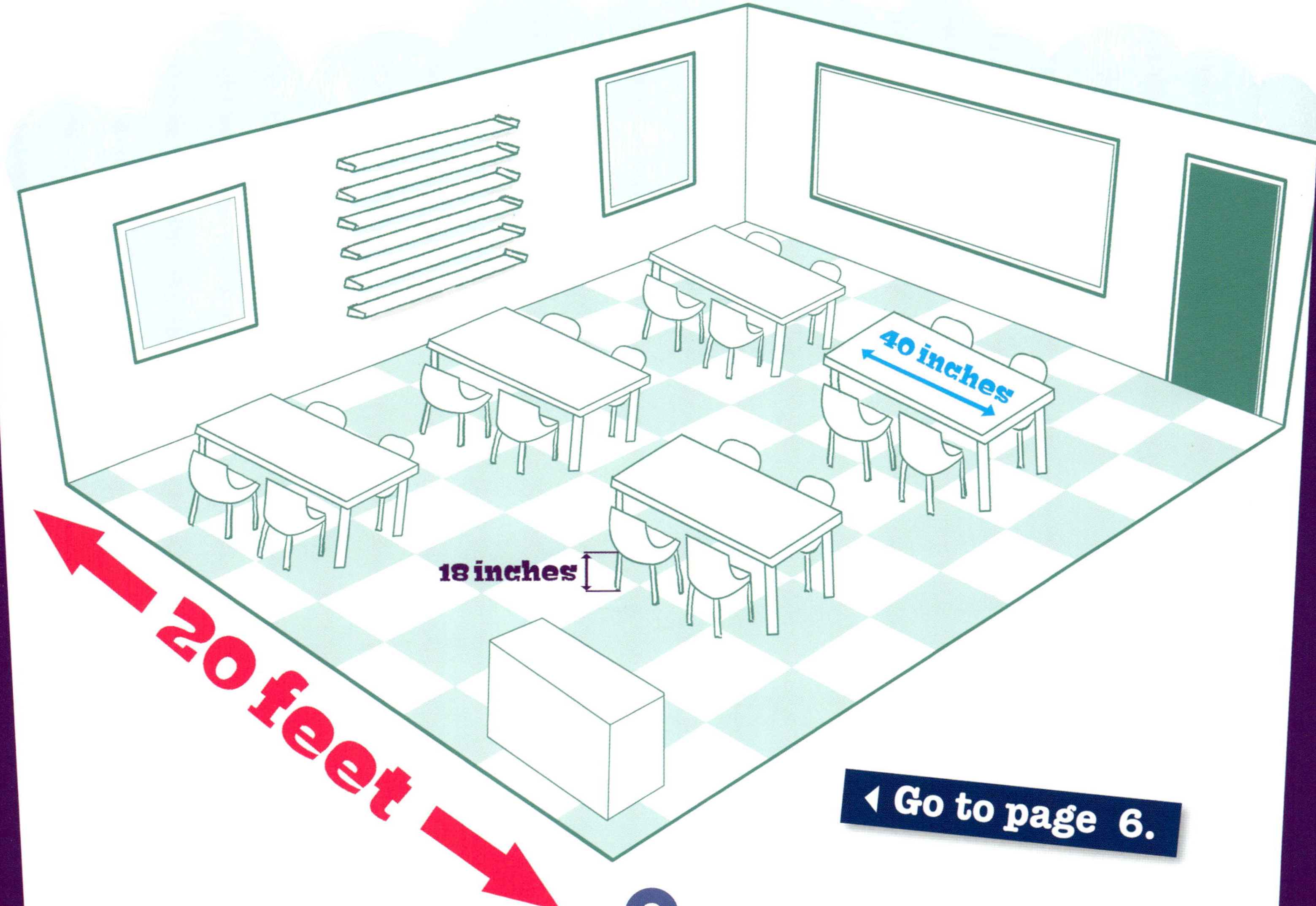

Go to page 6.

Think again.

It looks like the distance is more than one ruler. A ruler is one foot long. So Juan's ruler is too short. Try a yardstick instead.

Turn to page 15. ▸

“Good idea. We’ll cover the 10 feet on the right side of the room,” Juan says.

Ella finds a clue. “Look, cracker crumbs. Lucky loves crackers.”

“He can’t climb this smooth metal shelf though. Can he reach this high?” asks Juan.

"We can measure the distance, then compare that to Lucky's size," suggests Ella.

Juan says, "It looks like a little more than one foot. I estimate 15 inches."

"Should we measure it in inches or feet?" asks Mr. Jeb.

◀ **If they measure in inches, turn to page 6.**

◀ **If they measure in feet, turn to page 8.**

“Ella, you used centimeters,” Juan says. “Centimeters are a metric unit. They are smaller than inches. That’s why that number is so much larger.”

“Oops,” Ella says. She uses the inches side of her yardstick. “17 inches,” she says. “Lucky is only 9 inches. Even if he stretched, that’s still too high for him to reach.”

On the other side of the room, Mark finds another clue.

“Look at all these bits of yellow construction paper,” he says.

Turn to page 16.

Think again.

The room is 20 feet wide. Half of 20 is 10.

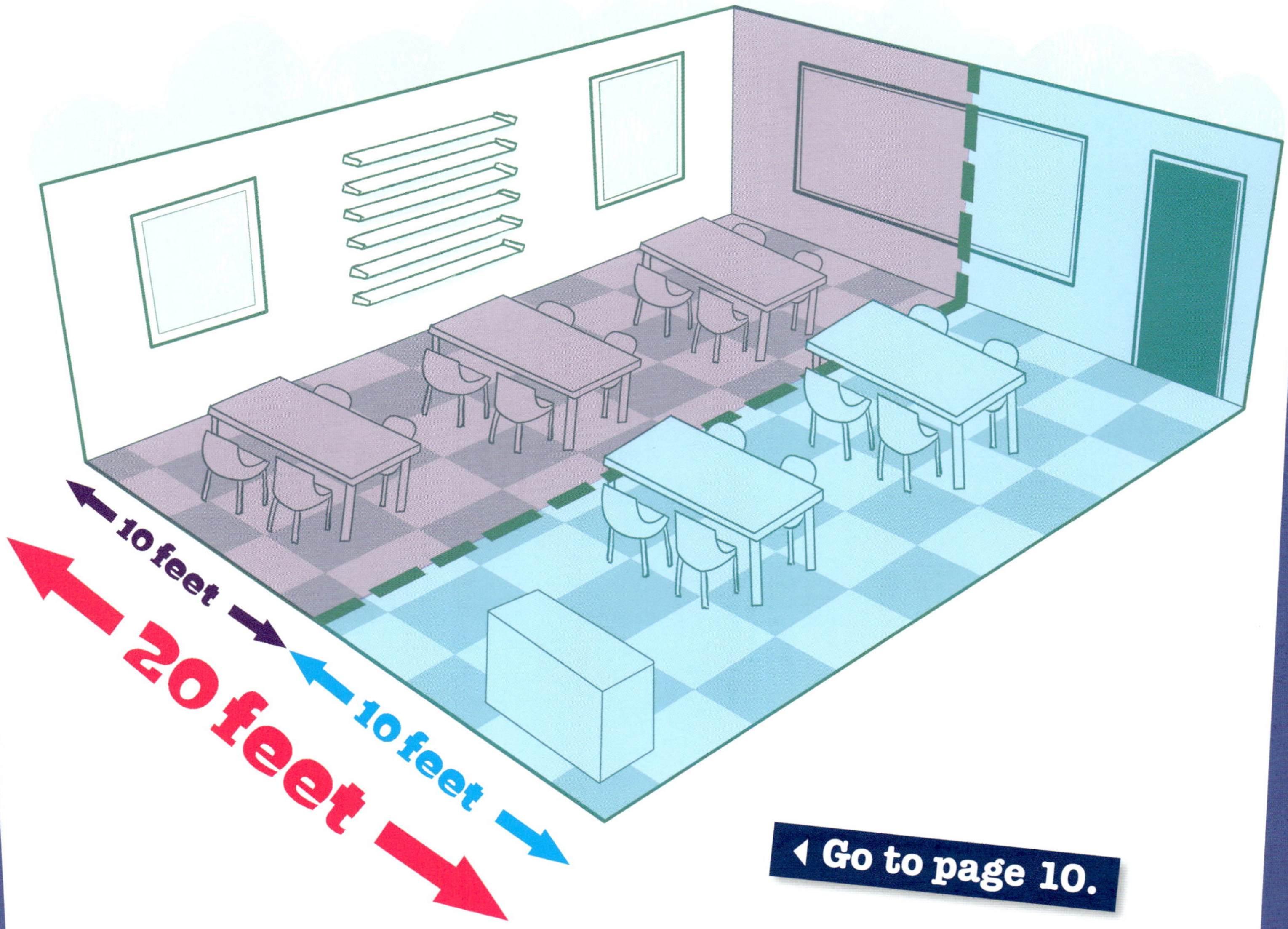

◀ Go to page 10.

Ella says, “I’ll use this yardstick. The shelf is 43 centimeters.”

They compare their measurement to Lucky’s size on his growth chart. Lucky is now 9 inches long.

Could Lucky have reached the shelf?

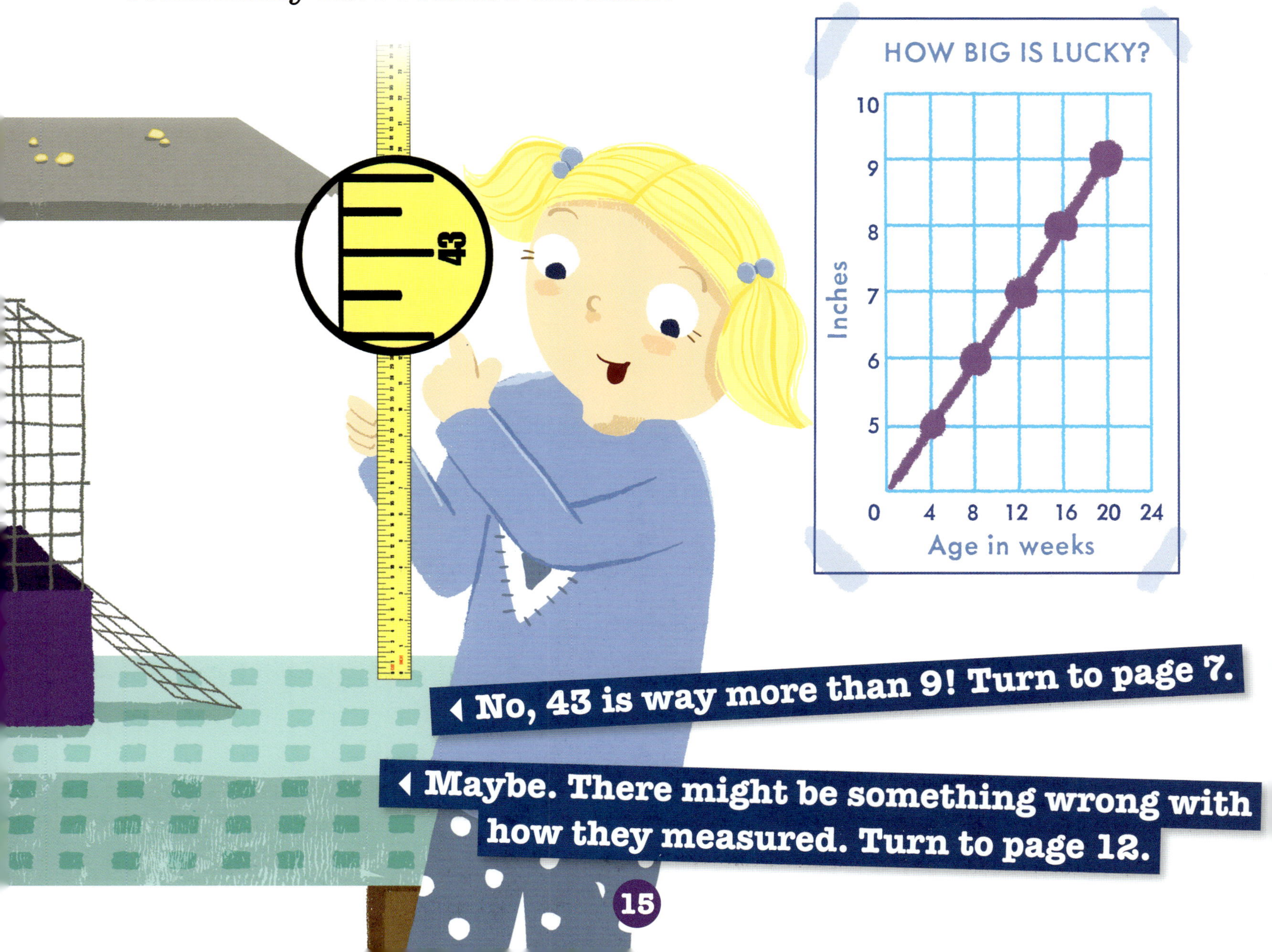

◂ **No, 43 is way more than 9! Turn to page 7.**

◂ **Maybe. There might be something wrong with how they measured. Turn to page 12.**

Jaden says, "There's more bits of yellow paper over here."

"There's more yellow paper up there, too," says Camila, pointing to a high bookshelf. "Maybe Lucky left a paper trail."

"He can sure climb these bins to get from shelf to shelf," Mark says.

"We'll need a ladder to check the top shelf. Let's measure to see what size ladder we need," Jaden suggests.

Turn to page 18. ▶

The other group joins them to help. Mark measures the shelves with the yardstick.

Ella says, "It's 3 feet to the third shelf. We can't reach to the top to measure all the way up."

Mr. Jeb says, "We don't have to reach all the way up. There are six shelves, all the same height."

Camila says, "We can just use math to measure the shelves."

◀ **If it is 7 feet to the top, go to page 5.**

If it is 6 feet to the top, go to page 20. ▶

"For every three shelves, we add three feet," says Camila. "So 3 feet plus 3 feet equals 6 feet."

Mr. Jeb says, "I'll get a ladder that can reach past 6 feet."

Mark adds, "And hurry! We need to find Lucky!"

$$\begin{array}{r} 3 \\ +3 \\ \hline 6 \end{array}$$

Mr. Jeb says, “I think he climbed in one of these art supply boxes. The paper trail leads to this one.” Soon, Lucky’s nose peeks out of the box.

Go to page 22. ▸

"Lucky made himself a little nest out of paper!" says Ella.
"Let's put him back in his cage to finish his nap," says Mr. Jeb.

"And make sure the door is latched!" Jaden says.
It's lucky Mr. Jeb's class knows how to measure.

Case, and cage, closed!

AMICUS ILLUSTRATED and **AMICUS INK**
are published by Amicus
P.O. Box 227, Mankato, MN 56002
www.amicuspublishing.us

Library of Congress Cataloging-in-Publication Data
Names: Adamson, Thomas K., 1970– author. | Adamson, Heather, 1974– author. | Alder, Charlie, illustrator.
Title: The case of the missing class pet / by Thomas K. and Heather Adamson ; illustrated by Charlie Alder.
Description: Mankato, Minnesota : Amicus, [2022]. | Series: Mini math mysteries | Audience: Ages 6–9. | Audience: Grades 2–3. | Summary: "Oh no! Lucky, the class pet, has escaped! Where did he go? Be a math detective along with the kids at Lincoln Elementary and help solve this pick-your-own-path mystery. Use math and measurement skills to figure out where Lucky could be hiding"—Provided by publisher.
Identifiers: LCCN 2019046016 (print) | LCCN 2019046017 (ebook) | ISBN 9781645490098 (library binding) | ISBN 9781681526515 (paperback) | ISBN 9781645490890 (pdf)
Subjects: LCSH: Area measurement—Juvenile literature. | Mathematical recreations—Juvenile literature. | Rats—Behavior—Juvenile literature.
Classification: LCC QC104.5 .A33 2022 (print) | LCC QC104.5 (ebook) | DDC 516/.156–dc23
LC record available at https://lccn.loc.gov/2019046016
LC ebook record available at https://lccn.loc.gov/2019046017

Editor: Rebecca Glaser
Designer: Kathleen Petelinsek

About the Authors

Thomas K. and Heather Adamson have both written many books for kids. This husband-and-wife team enjoys writing together. When they are not working, the couple likes to take hikes, watch movies, eat pizza, and of course, read. They live in South Dakota with their two sons and a Morkie named Moe.

About the Illustrator

Charlie Alder has written and illustrated many books for children, including *Daredevil Duck* and *Chicken Break!* She lives in Devon, England, with her husband and son. When not drawing chickens, Charlie can be found in her studio drinking coffee, arranging her crayons, and inventing more cheeky accidental heroes, both human and animal.